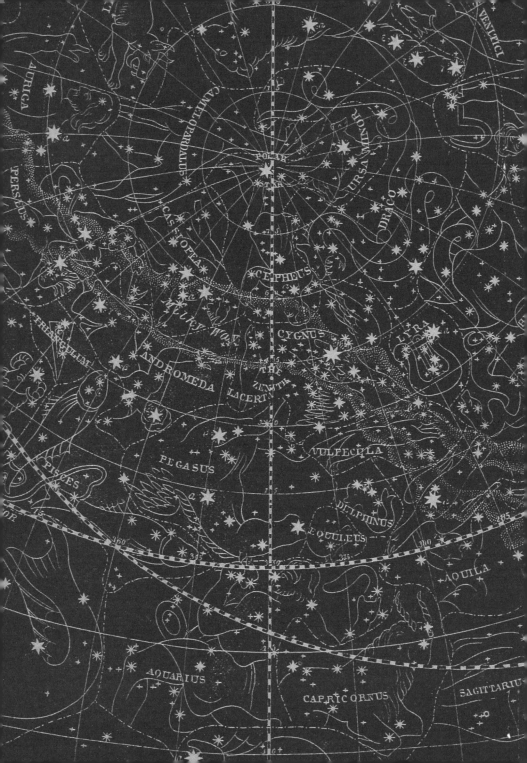

The Universe in Verse

The Universe in Verse

15 PORTALS TO
WONDER
THROUGH SCIENCE
& POETRY

Maria Popova

Illustrations by Ofra Amit

 Storey Publishing

♦

♦

♦

Science describes
accurately from outside,
poetry describes
accurately from inside.
Science explicates,
poetry implicates.
Both celebrate
what they describe.

—*Ursula K. Le Guin*

♦   ♦   ♦   *For Emily, who returned
her borrowed stardust
to the universe far too soon.*

Contents

POETRY,
SCIENCE, *and the*
COSMOS
of the POSSIBLE

We live our human lives in the
lacuna between truth and meaning,
between objective reality and
subjective sensemaking laced with
feeling. All of our longings, all of
our despairs, all of our reckonings
with the perplexity of existence
are aimed at one or the other. In the
aiming is what we call creativity,
how we contact beauty—the beauty
of a theorem, the beauty of a sonnet.

*T*he *Universe in Verse* was born in 2017 as a festival of wonder: stories from the history of science—the history of our search for truth and our yearning to know nature—told live onstage alongside readings of illustrative poems—those emblems of our search for meaning and our yearning to know ourselves. Year after year, thousands of people gathered to listen, think, and feel together—a congregation of creatures concerned with the relationship between truth and beauty, between love and mortality, between the finite and the infinite.

Poetry may seem an improbable portal into the fundamental nature of reality—into dark matter and the singularity, evolution and entropy, Hubble's law and pi—but it has a lovely way of sneaking ideas into our consciousness through the back door of feeling, bypassing our ordinary ways of seeing and relating to the world, our biases and preconceptions, and swinging open another gateway of receptivity. Through it, other scales of time, space, and significance—scales that are the raw material of science—can enter more fully and more faithfully into our worldview, depositing us back into our ordinary lives broadened and magnified so that we can return to our daily tasks and our existential longings with renewed resilience and a passion for possibility.

Poetry and science—individually, but especially together—are instruments for knowing the world more intimately and loving it more deeply. We need science to help us meet reality on its own terms, and we need poetry to help us broaden and deepen the terms on which we meet ourselves and each other.

At the crossing point of the two we may find a way of clarifying our experience and of sanctifying it; a way of harmonizing the objective reality of a universe insentient to our hopes and fears with the subjective reality of what it feels like to be alive, to tremble with grief, to be glad. Both are occupied with helping us discover something we did not know before—something about who we are and what this is. Their shared benediction is a wakefulness to reality aglow with wonder.

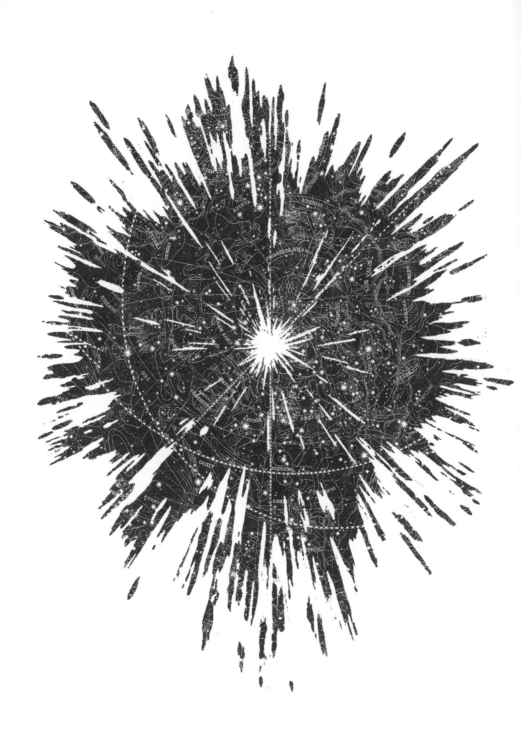

The Singularity
and Our
Elemental Belonging

*T*HE FIRST English use of the word *space* to connote the
cosmic expanse appears in line 650 of Book I of Milton's
Paradise Lost: "Space may produce new Worlds," he wrote,
and grow rife with them.

On this world, space has produced atoms with con-
sciousness, matter lustful of meaning. Minds. "The mind
is its own place," Milton wrote, "and in it self can make
a Heav'n of Hell, a Hell of Heav'n." Ours is a world rife
with minds—minds that have spent millennia seeking
to unravel the mysteries of space, making of this planet
heaven and hell in the process.

In the 1960s—a time of dueling theories about the origin
of the universe—the young physicist Stephen Hawking
had an idea: The spacetime singularities known to dwell
at the bottom of every black hole—points of zero radius
and infinite density, consuming and compressing every
datum that ever touches them under a pull of gravity so
intense that it ruptures the fabric of the universe—could
be applied to the whole of spacetime. Under such a model,

the entire universe would have arisen out of one such point—the Big Bang singularity.

It is an incomprehensible notion for the human mind to grasp—the human mind, itself made of atoms that were once part of particular stars, now constellating into molecules and cells firing electrical charges between neurons at eighty feet per second to produce the spark of thought, thought that cannot apprehend how everything we know, everything we love, everything that ever was and ever will be, was once compressed into a sizeless, senseless dot.

To come to terms with this elemental fact is to see ourselves—our inheritance and our interbelonging, our insignificance and our sanctity—with cleansed eyes, lensed with wonder and tenderness.

SINGULARITY

(after Stephen Hawking)

MARIE HOWE

Do you sometimes want to wake up to the singularity
we once were?

so compact nobody
needed a bed, or food or money—

nobody hiding in the school bathroom
or home alone

pulling open the drawer
where the pills are kept.

For every atom belonging to me as good
Belongs to you.   Remember?

There was no   *Nature.*   No
them.   No tests

to determine if the elephant
grieves her calf   or if

the coral reef feels pain.   Trashed
oceans don't speak English or Farsi or French;

~

would that we could wake up   to what we were
—when we *were* ocean   and before that

to when sky was earth, and animal was energy, and rock was
liquid and stars were space and space was not

at all—nothing

before we came to believe humans were so important
before this awful loneliness.

Can molecules recall it?
what once was?   before anything happened?

No I, no We, no one. No was
No verb   no noun
only a tiny tiny dot brimming with

is is is is is

All   everything   home

Flowers and
the Birth
of Ecology

Two hundred million years ago, long before we
walked the Earth, it was a world of cold-blooded
creatures and dull color—a kind of terrestrial sea of brown
and green. There were plants, but their reproduction was a
tenuous game of chance—they released their pollen into
the wind, into the water, against the staggering improbability
that it would reach another member of their species. No
algorithm, no swipe—just chance.

But then, in the Cretaceous period, flowers appeared
and carpeted the world with astonishing rapidity—because,
in some poetic sense, they invented love.

Once there were flowers, there was fruit—that tran-
scendent alchemy of sunlight into sugar. Once there was
fruit, plants could enlist the help of animals in a kind
of trade: sweetness for a lift to a mate. Animals savored the
sugars in fruit, converted them into energy and proteins,
and a new world of warm-blooded mammals came alive.

Without flowers, there would be no us.

No poetry.

No science.

No music.

Darwin could not comprehend how flowers could emerge so suddenly and take over so completely. He called it an "abominable mystery." But out of that mystery a new world was born, governed by greater complexity and interdependence and animal desire, with the bloom as its emblem of seduction.

In 1866, the young German marine biologist Ernst Haeckel—whose exquisite illustrations of single-celled underwater creatures had enchanted Darwin—gave that interdependence a name: He called it *ecology*, from the Greek *oikos*, or "house," and *logia*, "the study of," denoting the study of the relationships between organisms in the house of life.

A year earlier, in 1865, a young American poet—a keen observer of the house of life who made of it a temple of beauty—composed what is essentially a pre-ecological poem about ecology.

She had awakened to the interdependent splendor of the natural world as a teenager, when she composed a different kind of ecological poem: In a large album bound in green cloth, she painstakingly pressed, arranged, and labeled in her neat handwriting 424 wildflowers she had gathered from her native New England—some of them now endangered, some extinct.

This herbarium, which survives, became Emily Dickinson's first formal exercise in composition. Although she came to reverence the delicate interleavings of nature in many of her stunning, spare, strange, and always untitled poems, this one—the one she wrote in 1865, just before Ernst Haeckel coined *ecology*—illuminates and magnifies these relationships through the lens of a single flower and everything that goes into making its bloom—this emblem of seduction—possible: the worms in the soil (which Darwin celebrated as the unsung agriculturalists that shaped Earth as we know it), the pollinators in the spring air, all the creatures both competing for resources and symbiotically aiding each other.

And, suddenly, the flower emerges not as this pretty object to be admired, like it had been throughout the canon of Victorian poetry, but as this ravishing system of aliveness—a kind of silent symphony of interconnected resilience.

◆

◆

◆

[BLOOM]

EMILY DICKINSON

Bloom—is Result—to meet a Flower
And casually glance
Would cause one scarcely to suspect
The minor Circumstance

Assisting in the Bright Affair
So intricately done
Then offered as a Butterfly
To the Meridian—

To pack the Bud—oppose the Worm—
Obtain its right of Dew—
Adjust the Heat—elude the Wind—
Escape the prowling Bee

Great Nature not to disappoint
Awaiting Her that Day—
To be a Flower, is profound
Responsibility—

Entropy and the Art of Alternative Endings

I N 1865—a year before Ernst Haeckel coined the word
ecology—the German physicist Rudolf Clausius coined
the word *entropy* to describe the undoing of being. The
thermodynamic collapse of physical systems into increasing
levels of disorder and uncertainty. The dissolution of
cohesion along the arrow of time. Inescapable. Irreversible.
Perpetually inclining us toward, in poet Mary Ruefle's
perfect words, "the end of time, which is also the end of
poetry (and wheat and evil and insects and love)." Perpetually
ensuring, in poet Edna St. Vincent Millay's perfect words,
that "lovers and thinkers" become "one with the dull, the
indiscriminate dust."

This transformation of order into disorder, of constancy
into discontinuity, is how we register change and tell one
moment from the next. Without entropy, the universe would
be a vast eternal stillness—a frozen fixity in which never
and forever are one. Without entropy, there would be no
time—at least not for us, creatures of time.

Clausius built his new word on the Greek for "transformation," *tropē*, because he believed that leaning on ancient languages to name new scientific concepts made them available to all living tongues, belonging to all people for all time. It pleased him, too, that *entropy* looked like *energy*—its twin in the making and unmaking of the universe. Energy, the giver of life. Entropy, the taker away. The frayer of every cell that animates our bodies with being. The extinguisher of every star that unlooses its thermal energy into the cold sublime of spacetime as it runs out of fuel, warming up the orbiting planets with its dying breath. We are only alive because our Sun is burning out. Without entropy, there would be no us.

The child of a physicist, W. H. Auden had no illusion about the entropic nature of reality. He wove a science-lensed lucidity into his poetic search for truth, for meaning, for a way to live with our human fragility, with our twin capacities for terror and tenderness inside an impartial universe he knew to be impervious to our plans and pleas. A survivor of the two world wars, he had no illusion about how our humanity comes unwoven by its own pull but is also the enchanted loom that makes life worth living.

It was as Auden was reaching the peak of his poetic powers that the second war—the world's deadliest yet— broke out, brutal and incomprehensible. It may be that art is simply what we call our most constructive coping mechanism for the incomprehension of life and mortality, and so Auden coped through his art. He looked at the stars and saw "ironic points of light" above a world "defenseless under the night"; he looked at himself and saw a creature "composed like them of Eros and of dust, beleaguered by the same negation and despair."

His poem "September 1, 1939" became a generation's life raft for "the waves of anger and fear" subsuming the unexamined certainties of yore, splashing awake the "euphoric dream" of a final and permanent triumph over evil. But the war went on, and in the protracted post-traumatic reckoning with its aftermath—this gasping ellipsis in the narrative of humanity—Auden revised his understanding of the world, of life, of our human imperative, and so he revised his poem.

In what may be the single most poignant one-word alteration in the history of our species, he changed the final line of the penultimate stanza to reflect his war-annealed recognition that entropy dominates all. The original version reads: "We must love one another or die"— an impassioned plea for compassion as a moral imperative, the withholding of which assures the destruction of life. But the plea had gone unanswered and eighty million lives had gone unsaved. Auden came to feel that his reach for poetic truth had been rendered "a damned lie," later lamenting that however our ideals and idealisms may play out, "we must die anyway."

A decade of disquiet after the end of the war, he changed the line to read: "We must love one another and die."

But there was a private reckoning beneath the public one—this, after all, is the history of humanity, of our science and our art. Auden was working out the complexities of the world in the arena where we so often wrestle with the vastest, austerest, most abstract and universal questions about how reality works—the fleshy, feeling concreteness of personal love.

In the summer of 1939, just before the world came unworlded, Auden met the young aspiring poet Chester Kallman and fell in love, fell hard, fell dizzily into the strangeness of spending "the eleven happiest weeks" of his life amid a world haunted by death. Over the next two years, as the war peaked, this passionate love became a lifeline of sanity and survival. But Auden, already well into his thirties, kept longing for a stable and continuous relationship of mutual fidelity—the closest thing to a marriage their epoch allowed—while Kallman, barely twenty, kept wounding him with the scattered and discontinuous affections of self-discovery.

Throughout the cycles of heartache, Auden refused to withdraw his love—a stubborn and devoted love, opposing the forces of dissolution and disorder, outlasting the fraying of passion and the abrasions of romantic disappointment— until it buoyed their bond over to the other side of the tumult, to the stable shore of lifelong friendship.

For the remainder of his life, Auden summered with Kallman in Europe. They spent twenty New York winters as roommates in a second-floor apartment at 77 St. Mark's Place in the East Village, later marked with a stone plaque emblazoned with lines from Auden's ode to the foolish, fierce devotion that had prevailed over the lazy entropy of romantic passion to salvage from its wreckage the lasting friendship, the mutual cherishment and understanding that had bound them together in the first place.

THE MORE LOVING ONE

W. H. AUDEN

Looking up at the stars, I know quite well
That, for all they care, I can go to hell,
But on earth indifference is the least
We have to dread from man or beast.

How should we like it were stars to burn
With a passion for us we could not return?
If equal affection cannot be,
Let the more loving one be me.

Admirer as I think I am
Of stars that do not give a damn,
I cannot, now I see them, say
I missed one terribly all day.

Were all stars to disappear or die,
I should learn to look at an empty sky
And feel its total dark sublime,
Though this might take me a little time.

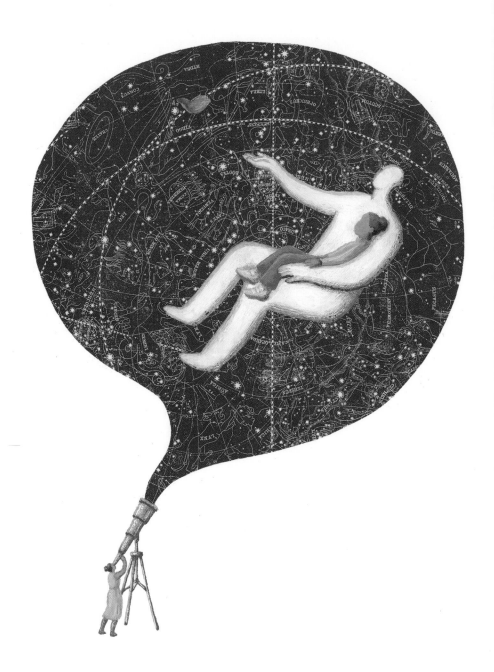

Henrietta Leavitt, Edwin Hubble, and Our Hunger to Know the Universe

*I*N 1908, Henrietta Swan Leavitt—one of the women known as the Harvard Computers, who changed our understanding of the universe long before they could vote—was analyzing photographic plates at the Harvard College Observatory, single-handedly measuring and cataloging more than two thousand variable stars—stars that pulsate like lighthouse beacons—when she began noticing a consistent correlation between their brightness and their blinking pattern. That correlation would allow astronomers to measure their distance for the first time, furnishing the yardstick of the cosmos.

Meanwhile, a teenage boy in the Midwest was repressing his childhood love of astronomy and beginning his legal studies to fulfill his dying father's demand for an ordinary, reputable life. Upon his father's death, Edwin Hubble would unleash his passion for the stars into formal study and lean on Leavitt's data to upend millennia of cosmic parochialism, demonstrating two revolutionary facts about the universe: that it is vastly bigger than we thought, and that it is growing bigger by the blink.

One October evening in 1923, perched at the foot of the world's most powerful telescope at Mount Wilson Observatory in California, Hubble took a 45-minute exposure of Andromeda, which was then thought to be one of many spiral nebulae in the Milky Way. The notion of a galaxy—a gravitationally bound swirl of stars and interstellar gas, dust, and dark matter—did not exist as such. The Milky Way—a name coined by Chaucer—was commonly considered an "island universe" of stars, beyond the edge of which lay cold dark nothingness.

When Hubble looked at the photograph the next morning and compared it to previous ones, he (I like to imagine) furrowed his brow. Then, with a gasp of revelation, he (this we know for a fact) crossed out the marking N on the plate, scribbled the letters V A R beneath it, and could not help adding an exclamation point.

Hubble had realized that a tiny fleck in Andromeda, previously mistaken for a nova, could not possibly be a nova, given its blinking pattern across the different photographs. It was a variable star—which, given Henrietta Leavitt's discovery, could only be so if the tiny fleck was very far away, farther than the edge of the Milky Way.

Andromeda was not a nebula in our own galaxy but a separate galaxy, out there in the cold dark nothingness.

Suddenly, the universe was a garden blooming with galaxies, with ours but a single bloom.

That same year, in another country suspended between two world wars, another young scientist, named

Hermann Oberth, was polishing the final physics on a daring idea: to subvert a deadly military technology with roots in medieval China and rocket-launch an enormous telescope into Earth's orbit—bringing it closer to the stars and bypassing the atmosphere that occludes our terrestrial instruments.

It would take two generations of scientists to make that telescope a reality—a shimmering poem of metal, physics, and perseverance, bearing Hubble's name.

But when the Hubble Space Telescope finally launched in 1990, hungry to capture the most intimate images of the cosmos humanity had yet seen, humanity had crept into the instrument's exquisite precision—its main mirror had been ground into the wrong spherical shape, warping its colossal eye.

Up the coast from Mount Wilson Observatory, a teenage girl watched her father—who had worked on the Hubble as one of NASA's first Black engineers—come home brokenhearted. He didn't know that his observant daughter would become poet laureate of his country and would come to commemorate him in the tenderest tribute an artist-daughter has ever made for a scientist-father. That tribute—the splendid poetry collection *Life on Mars*—earned Tracy K. Smith the Pulitzer Prize the year NASA published the revolutionary Ultra Deep Field image of the observable universe captured by the Hubble's corrected optics, revealing what neither Henrietta Leavitt nor Edwin Hubble could have imagined—that there isn't just one other galaxy besides our own, or just a handful more, but at least 100 billion, each containing at least 100 billion stars.

◆

◆

◆

from MY GOD,
IT'S FULL OF STARS

TRACY K. SMITH

When my father worked on the Hubble Telescope, he said
They operated like surgeons: scrubbed and sheathed
In papery green, the room a clean cold, a bright white.

He'd read Larry Niven at home, and drink scotch on the rocks,
His eyes exhausted and pink. These were the Reagan years,
When we lived with our finger on The Button and struggled

To view our enemies as children. My father spent whole seasons
Bowing before the oracle-eye, hungry for what it would find.
His face lit up whenever anyone asked, and his arms would rise

As if he were weightless, perfectly at ease in the never-ending
Night of space. On the ground, we tied postcards to balloons
For peace. Prince Charles married Lady Di. Rock Hudson died.

We learned new words for things. The decade changed.

The first few pictures came back blurred, and I felt ashamed
For all the cheerful engineers, my father and his tribe.
    The second time,
The optics jibed. We saw to the edge of all there is—

So brutal and alive it seemed to comprehend us back.

Dark Matter and Our Yearning for Light

M ONTHS BEFORE Edwin Hubble finally published
his epoch-making revelation about Andromeda,
staggering the world with the fact that the universe
extends beyond our Milky Way galaxy, a child was born
under the star-salted skies of Washington, D.C., where
the Milky Way was still visible before a century's smog
slipped between us and the cosmos—a child who would
grow up to confirm the existence of dark matter, that
invisible cosmic glue holding galaxies together and pinning
planets to their orbits so that, on at least one of them,
small awestruck creatures with vast complex conscious-
nesses can unravel the mysteries of the universe.

Night after night, Vera Rubin peered out of her
childhood bedroom and into the stars, wonder-smitten
with the beauty of it all—until she read a children's
book about the trailblazing astronomer Maria Mitchell,
who had expanded the universe of possibility for half of
our species a century earlier as America's first professional
female astronomer and the first woman employed by

the United States government for a "specialized non-domestic skill." After discovering a long-sought-after comet, she was hired as a "Computer of Venus" for the US Nautical Survey—a sort of one-woman GPS performing complex astronomical calculations to help sailors navigate the globe.

Reading about Maria Mitchell, the young Vera was suddenly seized with a life-altering realization: Not only was there such a thing as a professional stargazer, but it was a thing a girl could do.

In 1965—exactly one hundred years after Maria Mitchell was appointed the first professor of astronomy at Vassar and seventeen years after Vera Rubin graduated from it as an astronomer—she became the first woman permitted to use the Palomar Observatory. The telescope, devised the year Rubin was born, had replaced the one through which Hubble made his discovery as the world's most powerful astronomical instrument. Peering through its oracle eye, she was just as wonder-smitten as the little girl peering through the bedroom window, just as beguiled by the beauty of the cosmos. "I sometimes ask myself whether I would be studying galaxies if they were ugly," she reflected in her most personal interview. "I think it may not be irrelevant that galaxies are really very attractive."

Galaxies had taken Rubin to Palomar, and galaxies—the riddle of their rotation, which she had endeavored to solve—became the key to her epochal confirmation of dark matter. One of the most mesmerizing unsolved puzzles in astronomy, dark matter had remained only an enticing speculation since the Swiss astrophysicist Fritz

Zwicky had first theorized it when Vera was five. Dark matter may be the most urgent gap in our understanding of reality. It may be made of an undiscovered particle, or it may be something much more thrilling than a mere particle—there could exist dark protons and dark electrons visible to each other but invisible to us, fusing into dark atoms in an entire dark chemistry that gives rise to dark life existing all around us, imperceptible to us. And because dark matter comprises 95 percent of the matter in the universe, our visible life—our puny 5 percent—would be the oddity, the aberration, the ghostly alien on the fringe of the dark universe. Dark matter is not a mere gap in our understanding—it is the colossal contour of our ignorance.

A generation after Rubin's confirmation of the contour, a small clan of astronomers at Cambridge analyzed the deepest image of space the Hubble Space Telescope had yet captured—that iconic glimpse of the unknown, revealing a universe "so brutal and alive it seemed to comprehend us back"—to discern the origin of the mysterious dark matter halo enveloping the Milky Way. Spearheading the endeavor was an extraordinary young astronomer who had returned to work during a remission of a rare terminal blood cancer ordinarily afflicting the elderly.

Nursed on geology and paleontology on the shores of a prehistoric lake, Rebecca Elson was barely sixteen and already in college when she first glimpsed Andromeda through a telescope. Instantly dazzled by its "delicate wisp of milky spiral light floating in what seemed a bottomless well of empty space," she became a scientist but never relinquished the pull of the poetic dimensions of reality.

During her postdoctoral work at Princeton's Institute for Advanced Study, Elson found refuge from the narrow patriarchy of academic science in a gathering of poets every Tuesday evening. She became a fellow at a Radcliffe-Harvard institute for postgraduate researchers devoted to reversing "the climate of non-expectation for women," among the alumnae of which are Anne Sexton, Alice Walker, and Anna Deavere Smith. There, in a weekly writing group, she met and befriended the poet Marie Howe.

It was then—twenty-nine and newly elected the youngest astronomer in history to serve on the Decennial Review committee steering the course of American science toward the most compelling unsolved questions—that Elson received her terminal diagnosis.

Throughout the bodily brutality of her cancer treatment, she filled notebooks with poetic questions and experiments in verse, bridging with uncommon beauty the creaturely and the cosmic—those eternal mysteries of our mortal matter that make it impossible for a consciousness born of dead stars to fathom its own nonexistence.

Rebecca Elson lived with the mystery for another decade, never losing her keen awareness that we are matter capable of wonder, never ceasing to channel it in poetry. When she returned her borrowed stardust to the universe, a spring shy of her fortieth birthday, she left behind nearly sixty scientific papers and a trove of poems, posthumously published under the lovely title *A Responsibility to Awe*.

Permeating Elson's lyrical meditations on the nature of reality, mortised and tenoned with life and love, the mystery of dark matter culminates in one particular

poem exploring with uncommon loveliness what may be
the most touching paradox of being human—creatures
of matter in a cosmos governed by the dark sublime
of endless entropy, longing for the light of immortality,
longing to return to the singularity so that everything
may begin again.

◆

◆

◆

LET THERE ALWAYS BE LIGHT (SEARCHING FOR DARK MATTER)

REBECCA ELSON

For this we go out dark nights, searching
For the dimmest stars,
For signs of unseen things:

To weigh us down.
To stop the universe
From rushing on and on
Into its own beyond
Till it exhausts itself and lies down cold,
Its last star going out.

Whatever they turn out to be,
Let there be swarms of them,
Enough for immortality,
Always a star where we can warm ourselves.

Let there be enough to bring it back
From its own edges,
To bring us all so close we ignite
The bright spark of resurrection.

Emmy Noether, Symmetry, and the Hidden Order of Things

As he was revolutionizing our understanding of reality, Albert Einstein kept stumbling over one monolith of mystery—why it is that while some things in physical systems change (and relativity is a theory of change: of how changes in coordinates give shape to spacetime), nature keeps other things immutable: things like energy, momentum, and electrical charge. And the crucial puzzle: why we cannot destroy energy or create it out of nothing— we can only transform it from one form to another in ever-morphing symmetries.

The revelation, which solved a significant problem with Einstein's theory of general relativity, came from the mathematics of Emmy Noether.

Born into a Jewish family in rural Germany in 1882, the daughter of a mathematician, Emmy Noether showed an early and exquisite gift for mathematics: this abstract plaything of thought, this deepest language of reality.

She excelled through all the education available to her, completing her doctorate in 1907 as one of two women in a class of nearly a thousand, shortly after the government had declared that mixed-sex education would "overthrow all academic order."

For seven years, while Einstein was working out his theories, Noether was working without pay as a mathematics instructor at the local university. In 1915— the year Einstein's general relativity reframed our picture of reality—she finally received proper employment at the country's premier research institution. At the University of Göttingen, where three centuries of visionary scientists have honed their science and earned their Nobels, Emmy Noether developed the famous theorem now bearing her name. Considered one of the most important and beautiful in all of mathematics, it proves that conservation laws rely on symmetry.

A generation after the women decoding the universe for paltry pay at the Harvard College Observatory under the directorship of Edward Charles Pickering became known as "Pickering's Harem," Emmy Noether's mathematics students became known as the "Noether boys."

In 1932, she became the first woman to give the plenary address at the International Congress of Mathematicians—the world's most venerable gathering of brilliant abstract minds. Of the 420 participating mathematicians, Emmy Noether was the only woman. Another woman would not address the Congress until 1990—the year the Hubble Space Telescope leaned on her physics to open its colossal eye into the unseen cosmos. At the time I am writing this, three decades later, there has not been a third.

Months after Emmy Noether's address, the Nazis banished Jewish professors from German universities. The position she had spent half a century and a lifetime earning was vanquished overnight.

Einstein sought refuge in Princeton—that epicenter of physicists and mathematicians of his and her caliber. But Princeton had no room for a her. Emmy Noether ended up at Bryn Mawr. Although she was invited as a guest lecturer on the request of the working scientists at Princeton, whose field would have been unimaginable without her contribution, the university overlords made her feel unambiguously unwelcome. Even this cheerful and uncomplaining woman, too in love with the abstract beauty of mathematics to have been thwarted by the systemic exclusion of the body carrying the mind, rued that it was "the men's university, where nothing female is admitted."

Symmetry now permeates our understanding of the universe and the language of physics. It is nigh impossible to publish any paper—that is, to formulate any meaningful model of reality—without referring to symmetry in some way. This was Emmy Noether's gift to the world— a whole new way of seeing and a whole new vocabulary for naming what we see, which is the fundament of fathoming and sensemaking. What she gave us is not unlike poetry, which gives us a new way of comprehending what is already there but not yet noticed and not yet named. With her elegant, deeply original mathematics, which came to underpin the entire standard model of particle physics, Emmy Noether became the poet laureate of reality—a reality underpinned by hidden laws figured into its every littlest manifestation, mortised with the meaning we give them and give our lives.

FIGURES OF THOUGHT

HOWARD NEMEROV

To lay the logarithmic spiral on
Sea-shell and leaf alike, and see it fit,
To watch the same idea work itself out
In the fighter pilot's steepening, tightening turn
Onto his target, setting up the kill,
And in the flight of certain wall-eyed bugs
Who cannot see to fly straight into death
But have to cast their sidelong glance at it
And come but cranking to the candle's flame—

How secret that is, and how privileged
One feels to find the same necessity
Ciphered in forms diverse and otherwise
Without kinship—that is the beautiful
In Nature as in art, not obvious,
Not inaccessible, but just between.

It may diminish some our dry delight
To wonder if everything we are and do
Lies subject to some little law like that;
Hidden in nature, but not deeply so.

Trees and the Optimism of Resilience

 REES GRANT US some of the richest metaphors for our
own lives—a polished lens on the quality of attention
we pay the world. "The tree which moves some to tears of
joy is in the eyes of others only a green thing which stands
in the way," wrote William Blake. Walt Whitman considered
them our greatest teachers in living with authenticity. For
Hermann Hesse, the key to existential joy was in learning
how to listen to the trees.

But far beyond the realm of human-wrested metaphor,
trees are sovereign marvels of nature, dazzling in the native
poetics of their biological and ecological reality. Their
photosynthesis is nature's way of making life from light.
Chlorophyll—which shares a chemical kinship with the
hemoglobin in our blood—allows a tree to capture photons,
extracting a portion of their energy to make the sugars
that make it a tree—the raw material for leaves and bark
and roots and branches—then releasing the photons at
lower wavelengths back into the atmosphere. A tree is a
light-catcher that grows life from air—an enormous eye
tuned to the light of the universe.

Trees hungrily absorb red light—the longer wavelengths of the visible spectrum—but the neighboring infrared passes straight through them. Under the canopy, where fierce competition for these wavelengths rages, red light is depleted and infrared dominates. Even though trees cannot absorb infrared, they, unlike humans, can "see" it with chemical photoreceptors called phytochromes. The ratio between the two types of light tells trees how much to grow and in which direction, with phytochromes acting as on-off switches for growth. An abundance of red light under uncrowded skies turns the switch on, signaling to the tree to spread its branches wide into any gaps in the canopy; in the crowded shade where infrared dominates, the switch turns off, reducing the growth of side branches and prompting the tree to grow straight up, reaching for the open sky above.

As summer recedes into autumn, cooling the air and dimming the light, the alchemy of transmuting light into growth becomes too metabolically costly for deciduous trees. Chlorophyll begins to break down, revealing the other pigments that had been there all along—the yellow of xanthophyll, the orange of carotenoids, the reds and purples of anthocyanins, turning the canopy into an aria of color.

Meanwhile, the layer of cells by which the stem holds on to the branch is fraying. Leaves begin to let go—a process known as abscission. But as they denude the branches, they reveal the subtle nubs of the new buds that had been forming all summer, readying next spring's growth. Skeletal and pulmonary, winter trees rise into the leaden sky, their skin a braille poem of resilience.

OPTIMISM

JANE HIRSHFIELD

More and more I have come to admire resilience.
Not the simple resistance of a pillow, whose foam
returns over and over to the same shape, but the sinuous
tenacity of a tree: finding the light newly blocked on one side,
it turns in another. A blind intelligence, true.
But out of such persistence arose turtles, rivers,
mitochondria, figs—all this resinous, unretractable earth.

Pi and the Seductions of Infinity

"MY BUSINESS IS CIRCUMFERENCE," Emily Dickinson wrote in one of her most cryptic letters. Since ancient times, human beings have been enchanted by the immutable relationship between the circumference of a circle and its diameter, no matter the circle's size. Today, we have a symbolic form for that mathematical relationship: π—an ancient Greek letter conferred upon it by a Welsh mathematician in the first years of the eighteenth century, though it was the ancient Greeks themselves who first began thinking mathematically about the mysterious number. The longest number in nature and possibly the most powerful, π factors into our understanding of fractals and eclipses, of cosmology and thermodynamics, yet it remains ever elusive in its totality.

In the third century BCE, a millennium after Babylonian and Egyptian scholars tried to discern its exact value with fractions, Archimedes devised a geometrical approach that contoured its first few digits. Eight centuries later, ancient Chinese and Indian mathematicians approximated it to seven digits. The invention of calculus in the seventeenth century bloomed hundreds of digits, with Newton himself computing the first fifteen. Modern supercomputers can calculate with perfect precision 1.4 trillion digits. We need only the first thirty-two to compute the size of the known universe with a margin of error a single proton wide.

An irrational number—a number that cannot be expressed as a fraction, the ratio between two whole numbers—π unmoors our basic intuitions about reality with its disquieting whisper of an infinity beyond the grasp of reason. There are no known infinities in nature— as transient creatures suspended in space between the scale of atoms and the scale of stars, suspended in time between not yet and no more, we simply cannot conceive of infinity. And yet the decimal point of π taunts us like the gun barrel of the unimaginable. If we ever reach the last digit of π, we will have known the universe. Meanwhile, its assuring constancy goes hedging against our own transience, slaking our yearning for permanence in a cosmos governed by incessant change.

P I

WISŁAWA SZYMBORSKA

translated by Clare Cavanagh and Stanisław Barańczak

The admirable number pi:
three point one four one.
All the following digits are also initial,
five nine two because it never ends.
It can't be comprehended *six five three five* at a glance,
eight nine by calculation,
seven nine or imagination,
not even *three two three eight* by wit, that is, by comparison
four six to anything else
two six four three in the world.
The longest snake on earth calls it quits at about forty feet.
Likewise, snakes of myth and legend, though they may
     hold out a bit longer.
The pageant of digits comprising the number pi
doesn't stop at the page's edge.
It goes on across the table, through the air,
over a wall, a leaf, a bird's nest, clouds, straight into the sky,
through all the bottomless, bloated heavens.
Oh how brief—a mouse tail, a pigtail—is the tail of a comet!
How feeble the star's ray, bent by bumping up against space!
While here we have *two three fifteen three hundred nineteen*
my phone number your shirt size the year
nineteen hundred and seventy-three the sixth floor

～

◆

◆

◆

the number of inhabitants sixty-five cents
hip measurement two fingers a charade, a code,
in which we find *hail to thee, blithe spirit, bird thou never wert*
alongside *ladies and gentlemen, no cause for alarm,*
as well as *heaven and earth shall pass away,*
but not the number pi, oh no, nothing doing,
it keeps right on with its rather remarkable *five,*
its uncommonly fine *eight,*
its far from final *seven,*
nudging, always nudging a sluggish eternity
to continue.

Euclid and the Dazzling Beauty of Universal Truth

*T*WENTY-FOUR CENTURIES AGO, with nothing but
an unmarked ruler and a compass, Euclid laid the
foundation of geometry by wresting from nature a set of
postulates that gave reality its first foothold of stability—
self-evident truths that held up across time and space,
dazzling in their austere beauty, infinitely testable and
provable, existing deeper and higher than faith.

In perfect Euclidean geometry, the angles inside
a triangle always add up to 180 degrees—the very first
mathematical proof, a validation of truth unmoored
from human judgment and opinion. It was the lightning
bolt that sundered the tree of knowledge into philosophy
and pure science. Its axiomatic splendor became a model
for clear thinking that shaped the development of logic.
Neither Kepler nor Galileo, neither Newton nor Einstein,
could have done their own revolutionary work without
leaning on Euclid.

After surviving for millennia on papyrus, Euclid's
Elements became second only to the Bible in the number of

editions published since the invention of the printing press. Lincoln would study his copy by lamplight and carry it in his saddlebag. Einstein would come to regard his own as his "holy little geometry book." Emerson would celebrate Euclid as a poet on par with Dante.

Edna St. Vincent Millay was twenty-one when she enrolled in Vassar College as a freshman in 1913. Immersed in the strong science curriculum established there by pioneering astronomer Maria Mitchell, who had paved the way for women in science in the previous century, the young poet found in geometry a singular portal to truth and beauty. The principles of geometry struck her as "so pure, so relentless and incorruptible"—like Bach. She composed one of her earliest sonnets—a paean of resistance to the religious mythologies purveying dogmas to be taken on faith—as a tribute to Euclid's legacy of revolutionizing how we look at the world and what we see, celebrating his eternal invitation to relish the beauty in reality laid bare.

EUCLID ALONE HAS LOOKED ON BEAUTY BARE

EDNA ST. VINCENT MILLAY

Euclid alone has looked on Beauty bare.
Let all who prate of Beauty hold their peace,
And lay them prone upon the earth and cease
To ponder on themselves, the while they stare
At nothing, intricately drawn nowhere
In shapes of shifting lineage; let geese
Gabble and hiss, but heroes seek release
From dusty bondage into luminous air.
O blinding hour, O holy, terrible day,
When first the shaft into his vision shone
Of light anatomized! Euclid alone
Has looked on Beauty bare. Fortunate they
Who, though once only and then but far away,
Have heard her massive sandal set on stone.

Radioactivity and the Mystery of Matter

THE ANCIENT GREEKS were the first to seek an organizing principle for what the world is made of. For them, substances were its fundamental units of matter. They were called elements and there were four and you could feel them on your skin, run your fingers through them and test them with your feet. Democritus had theorized the atom, only to be laughed into two millennia of oblivion, until chemistry conceived of atomic theory in the late nineteenth century and gave birth to particle physics in the early twentieth. Then substances suddenly came undone to reveal smaller fundamental units of matter, which too were called elements, and there were dozens of them, then a hundred, then more, endlessly combinable and recombinable into everything we can see, smell, touch, taste, and turn into a wedding band or a bomb.

And then, in 1902, something unthinkable happened: A New Zealand physicist and his English colleague watched the central certainty of physics come undone in their Montreal laboratory as one element became another—something a French physicist had theorized six years earlier while bringing to his longtime fascination with phosphorescence the miracle of X-rays, just discovered by a German physicist.

Suddenly, matter was mutable and its organizing principle was not that of substances, nor of elements, but of the real fundamental units—protons and electrons, endlessly configurable, capable of what the alchemists had failed at: transmuting one element into another, simply by shedding a particle.

Within a couple of years, in a Parisian laboratory, a solemn Polish physicist and chemist in a radium-stained black dress discovered two new radioactive elements and turned uranium into the most precise timepiece humanity has ever known, suddenly revealing the age of rocks and the age of fossils—the true geologic genealogy of our habitable world and the evolutionary history of its inhabitants.

It wasn't then known that radioactive elements frayed the living structure of the human body. As her discoveries earned her one Nobel Prize, then another, Marie Curie— the first woman awarded the esteemed accolade, and to this day the only scientist awarded it in two different sciences, chemistry and physics—went on handling the vials of radium with her bare hands. Drawing on her research, she invented an X-ray ambulance that, together with her daughter, she chauffeured through the wreckage of World War I, saving countless lives, not knowing she herself was dying.

POWER

ADRIENNE RICH

Living   in the earth-deposits   of our history

Today a backhoe divulged   out of a crumbling flank of earth
one bottle   amber   perfect   a hundred-year-old
cure for fever   or melancholy   a tonic
for living on this earth   in the winters of this climate

Today I was reading about Marie Curie:
she must have known she suffered   from radiation sickness
her body bombarded for years   by the element
she had purified
It seems she denied to the end
the source of the cataracts on her eyes
the cracked and suppurating skin   of her finger-ends
till she could no longer hold   a test-tube or a pencil

She died   a famous woman   denying
her wounds
denying
her wounds   came   from the same source as her power

The Octopus
and the
Unknown

To LIVE wonder-smitten with reality is the gladdest
way to live. But with our creaturely capacity for wonder
comes a responsibility to it—the recognition that reality
is not a singularity but a plane. Each time we presume
to have seen the whole, the plane tilts ever so slightly to
reveal new vistas of truth and new horizons of mystery,
staggering us with the sudden sense that we had been
looking at only a fragment, framed by our parochial
point of view. The history of our species is the history of
learning and forgetting and relearning this elemental truth.

For eons, we assumed that Earth was the center of
the universe, all other celestial bodies revolving around it.
For eons, we assumed that the human animal was the
center of nature, all other life-forms revolving around it;
we assumed that we alone have consciousness. Descartes,
who so greatly advanced the scientific method forward
by pioneering empiricism, also paralyzed our understanding
of the mind with his dogmatic declamation that non-
human animals are automata—fleshy robots governed by

mechanistic reflexes, insentient and incapable of feeling. It took four centuries for this dogma to be upended after a young primatologist began her paradigm-shifting work in Gombe National Forest in 1960. Despite the tidal wave of dismissal and derision from the scientific establishment, Jane Goodall persisted, revolutionizing our understanding of consciousness and of our place in the family of life.

More than half a century later, some of the world's leading neuroscientists composed and cosigned the Cambridge Declaration on Consciousness, asserting that a vast array of nonhuman animals are also endowed with consciousness. The list named only one invertebrate.

The octopus branched from our shared vertebrate lineage some 550 million years ago to evolve into one of this planet's most alien intelligences, endowed with an astonishing distributed nervous system and capable of recognizing others, of forming social bonds, of navigating mazes. It is the Descartes of the oceans, learning how to live in its environment by trial and error—that is, by basic empiricism. For eons, the most majestic members of the octopus family, dwellers of the darkest depths, were thought to be impossible—until a pioneering deep-sea expedition in the final years of the nineteenth century observed creatures never before seen or imagined, upending the long-held assumption that life could not exist below 300 fathoms, shifting the lens to reveal yet another vista of reality, yet another plane of the possible.

IMPOSSIBLE BLUES

MARIA POPOVA

There,
at the bottom of being,
where the water that makes
   this planet a world
      is the color of spacetime

the octopus—

with her body-shaped mind
and her eight-arm embrace
   of alien realities,
with her colorblind vision
   sightful of polarized light,
and her perpetually awestruck
   lidless eye—
      can see

shades of blue we cannot conceive.

Call it god
   if you must
lean on the homely
to fathom the holiness
   of the fathomless whole.

~

And meanwhile,
up here,
    we swim amid particles
we cannot perceive
    folded into dimensions
we cannot imagine

to tell stories about
    what is real and
    what is possible,
and what it means to be.

A blink of time ago
we thought this octopus
    impossible,
this blue world
    lifeless
        below three hundred fathoms
until
an epoch after Bach
    scribbled in the margin
        of a composition
            "Everything that is possible is real"—
we plunged our prosthetic eye
    deep into the blue
and found a universe of life.

There,
the octopus,
godless and possible,
lives.

This Mote
of Matter

IN THE SUMMER OF 1977, NASA launched the *Voyager 1* and *Voyager 2* spacecrafts into the cosmos—twin poems of aluminum and electricity, tasked with exploring and documenting the outer planets of our Solar System on their way into the great beyond. As the *Voyagers* sailed away from Earth at thirty-five thousand miles an hour, they took spectacular technicolor images of the outer planets—the first-ever complete portrait of our cosmic neighborhood. They saw for the first time, in rapturous detail, the Jovian moons Galileo had glimpsed as faint spots and named after Roman deities nearly four centuries earlier. They saw Uranus, the color of my father's eyes, its aquamarine so arresting and thoroughly unexpected that it extracted audible gasps from the imaging team on the ground. They saw what no human imagination had fathomed before: Navigation engineer Linda Morabito, glancing back over the shoulder of the spacecraft to take one final photograph of Jupiter, was stunned by the sight of an enormous umbrella-shaped plume emerging from behind its moon Io. In an instant, she was seized by the certainty that she was seeing the first known evidence of active volcanism on a world other than Earth—a Vesuvius four hundred million miles from home.

Within four decades, one spacecraft would overtake the other and become the first human-made object to break out of the Sun's magnetic field and enter the enigma of interstellar space.

Neptune was the final target of humanity's enormous prosthetic eye. Planetary scientist Heidi Hammel—a meticulous marginalian in her mission logs—captured the rapture of its striking blue orb, so kindred to Earth yet so otherworldly, in three letters and a symbol in her log notes: "WOW!"

With the imaging mission complete after Neptune, NASA commanded that the cameras be shut down to conserve energy. But the astronomer Carl Sagan had the simple, revolutionary idea to turn the camera around and take one final photograph—of Earth itself. Objections were raised—from so great a distance and at so low a resolution, there would be absolutely no scientific value to the image. But Sagan, convinced of the photo's poetic worth, appealed all the way to the top and persuaded the administrator of NASA to grant permission. I can only imagine the words this poet laureate of the cosmos chose in making his case for the poetic over the purely scientific.

On Valentine's Day 1990, as the silent serpent of cancer was weaving its way through Sagan's body undetected, the *Voyager* pointed its cameras toward the inner Solar System from its distant vantage point 3.7 billion miles away. The other planets, while faint, were easy to discern. But Earth seemed to have vanished in the light-streaked photograph— a blurry speck islanded in a stream of sunlight against the blackness of empty space. There amid the endless expanse of nowheres, amid the shoreless cosmic ocean of pure spacetime that floods the vast majority of the universe, was our tiny somewhere. A mere 12 percent of a single pixel on a 640,000-pixel image.

This landmark photograph would come to be known as the *Pale Blue Dot*, after a phrase from the beautiful speech about the *Voyager* mission Sagan delivered at Cornell University in 1994—a speech that swelled into a best-selling book the following year, reverencing Earth as a "mote of dust suspended in a sunbeam," on which "everyone you love, everyone you know, everyone you ever heard of, every human being who ever was, lived out their lives."

Meanwhile, one of Earth's greatest poets was listening. In 1995, to celebrate the fiftieth anniversary of the United Nations, Maya Angelou composed a clarion call to humanity that would soar into space aboard NASA's *Orion* spacecraft. She dedicated the poem to "the hope for peace, which lies, sometimes hidden, in every heart." At its climax, she uses the phrase "mote of matter"—a subtle tribute to Carl Sagan and the message of the *Voyager*, a message about our place in the cosmic order not as something separate from and superior to nature, but as a tiny pixel-part of it, imbued with equal parts humility and responsibility.

◆

◆

◆

A BRAVE AND STARTLING TRUTH

MAYA ANGELOU

We, this people, on a small and lonely planet
Traveling through casual space
Past aloof stars, across the way of indifferent suns
To a destination where all signs tell us
It is possible and imperative that we learn
A brave and startling truth

And when we come to it
To the day of peacemaking
When we release our fingers
From fists of hostility
And allow the pure air to cool our palms

When we come to it
When the curtain falls on the minstrel show of hate
And faces sooted with scorn are scrubbed clean
When battlefields and coliseum
No longer rake our unique and particular sons and daughters
Up with the bruised and bloody grass
To lie in identical plots in foreign soil

~

When the rapacious storming of the churches
The screaming racket in the temples have ceased
When the pennants are waving gaily
When the banners of the world tremble
Stoutly in the good, clean breeze

When we come to it
When we let the rifles fall from our shoulders
And children dress their dolls in flags of truce
When land mines of death have been removed
And the aged can walk into evenings of peace
When religious ritual is not perfumed
By the incense of burning flesh
And childhood dreams are not kicked awake
By nightmares of abuse

When we come to it
Then we will confess that not the Pyramids
With their stones set in mysterious perfection
Nor the Gardens of Babylon
Hanging as eternal beauty
In our collective memory
Not the Grand Canyon
Kindled into delicious color
By Western sunsets

Nor the Danube, flowing its blue soul into Europe
Not the sacred peak of Mount Fuji
Stretching to the Rising Sun
Neither Father Amazon nor Mother Mississippi who,
    without favor,
Nurture all creatures in the depths and on the shores
These are not the only wonders of the world

When we come to it
We, this people, on this minuscule and kithless globe
Who reach daily for the bomb, the blade and the dagger
Yet who petition in the dark for tokens of peace
We, this people on this mote of matter
In whose mouths abide cankerous words
Which challenge our very existence
Yet out of those same mouths
Come songs of such exquisite sweetness
That the heart falters in its labor
And the body is quieted into awe

We, this people, on this small and drifting planet
Whose hands can strike with such abandon
That in a twinkling, life is sapped from the living
Yet those same hands can touch with such healing,
    irresistible tenderness
That the haughty neck is happy to bow
And the proud back is glad to bend
Out of such chaos, of such contradiction
We learn that we are neither devils nor divines

~

When we come to it
We, this people, on this wayward, floating body
Created on this earth, of this earth
Have the power to fashion for this earth
A climate where every man and every woman
Can live freely without sanctimonious piety
Without crippling fear

When we come to it
We must confess that we are the possible
We are the miraculous, the true wonder of this world
That is when, and only when
We come to it.

The Search
for Life

*I*N REFLECTING on the legacy of the Golden Record—
the disc encoded with music and images representing
humanity that sailed into the distant reaches of the cosmos
aboard the *Voyager*—Carl Sagan depicted us as "a species
endowed with hope and perseverance, at least a little
intelligence, substantial generosity and a palpable zest to
make contact with the cosmos."

Seven years later, the year I was born, astronomer
Jill Tarter cofounded SETI—an institute dedicated to the
search for extraterrestrial intelligence—and Sagan, who
was an ardent supporter of the SETI project, began writing
his novel *Contact*. Published in 1985, it was adapted into
a major motion picture twelve years later, starring Jodie
Foster as an astronomer modeled on Dr. Tarter. In the
most stirring scene, Foster's character peers out her space-
ship window as she approaches an extraordinary alien
world and gasps:

"They should've sent a poet!"

But the seed of that lyrical sentiment was planted in
Sagan much earlier, more than a decade before SETI was

born. In the early 1970s, he found himself so enchanted with the work of a young Cornell poet writing scientifically accurate poems about the universe—one of which Sagan sent to his friend Timothy Leary in prison—that he offered to be her doctoral advisor.

Diane Ackerman went on to compose a lifetime of poems and lyrical prose reverencing our relationship to the universe. "We Are Listening," inspired by the radio telescopes of SETI, now reads like the anthem for a new generation of searchers for life beyond Earth as our space telescopes go on discovering thousands of potentially habitable worlds beyond the edge of our Solar System, carrying our dreams of contact into the cosmos.

WE ARE LISTENING

DIANE ACKERMAN

I.

As our metal eyes wake
to absolute night,
where whispers fly
from the beginning of time,
we cup our ears to the heavens.
We are listening

on the volcanic lips of Flagstaff
and in the fields beyond Boston,
in a great array that blooms
like coral from the desert floor,
on highwire webs patrolled
by computer spiders in Puerto Rico.

We are listening for a sound
beyond us, beyond sound,

searching for a lighthouse
in the breakwaters of our uncertainty,
an electronic murmur
a bright, fragile *I am*.

~

Small as tree frogs
staking out one end
of an endless swamp,
we are listening
through the longest night
we imagine, which dawns
between the life and time of stars.

II.

Our voice trembles
with its own electric,
we who mood like iguanas
we who breathe sleep
for a third of our lives,
we who heat food
to the steaminess of fresh prey,
then feast with such baroque
good manners it grows cold.

In mind gardens
and on real verandas
we are listening,
rapt among the Persian lilacs
and the crickets,
while radio telescopes
roll their heads, as if in anguish.

With our scurrying minds
and our lidless will
and our lank, floppy bodies
and our galloping yens
and our deep, cosmic loneliness
and our starboard hearts
where love careens,
we are listening,
the small bipeds
with the giant dreams.

Mushrooms and the Creative Spirit

*T*HEY WERE THE FIRST to colonize the Earth. They will inherit it long after we are gone as a species. And when we go as individuals, it is they who return our borrowed stardust to the universe, feasting on our mortal flesh to turn it into oak and blackbird, grass and grasshopper. They undergird the forest in what is now known as the "wood wide web"—the mycorrhizal network across which root systems exchange nutrients and communicate in a language of sugar molecules and electrical signals. Each cubic inch of mycelium compresses eight miles of fine filaments folded unto themselves—the original superstrings of this terrestrial universe.

Fungi are the mightiest kingdom of life, and the least understood by our science, and the most everlasting. Without them, this planet would not be a world. Like everything vast and various, they shimmer with metaphors for life itself.

Sylvia Plath was twenty-seven and pregnant with her first child, a daughter, when she wrested from mushrooms a strange and poignant metaphor for the creative spirit.

In the setting summer in 1959, Plath and her complicated husband, Ted Hughes, arrived at Yaddo—the gilded artist's colony in Saratoga Springs, New York—and took up separate residences a five-minute walk apart. She had her first room of her own—a sunny third-floor studio in one of the larger houses, with a heavy wooden writing table and a hospital-green portable Swiss typewriter. Perched at her window, she watched the thicket of pines and listened to the birds. "I have never in my life felt so peaceful and as if I can read and think and write for about 7 hours a day," she wrote to her mother.

But it was also a season of dejection: One of America's oldest and largest publishers had just rejected her poetry manuscript, another rejected her first children's book—a story about living free from the world's estimation—and her depression crept back after a pleasantly distracting summer road trip.

She sorrowed in her journal:

> *Very depressed today. Unable to write a thing.*
> *Menacing gods. I feel outcast on a cold star,*
> *unable to feel anything but an awful helpless*
> *numbness . . . Caught between the hope and*
> *promise of my work . . . and the hopeless gap*
> *between that promise, and the real world of*
> *other people's poems and stories and novels.*

In the evenings, Ted and some of the other residents engaged in antiscientific entertainment—astrology charts and Ouija boards. She participated without enthusiasm, perhaps because she had been spending her days devotedly studying German—the language of rationalism and the Golden Age of Science, of Kant and Humboldt. By early November, she was seized with total creative block:

> *Paralysis again. How I waste my days. I feel a*
> *terrific blocking and chilling go through me*
> *like anesthesia . . . If I can't build up pleasures*
> *in myself: seeing and learning about painting,*
> *old civilizations, birds, trees, flowers, French,*
> *German . . . To give myself respect, I should*
> *study botany, birds and trees: get little booklets*
> *and learn them, walk out in the world.*

Walk out she did. The woods around Yaddo were damp and rife with mushrooms. Mushrooms were in the lavish food served at the colony. Mushrooms crept onto her mind.

And then, in that way that noticing has of revivifying the deadened spirit, she started to come alive, as if assured by nature that life—like fungi, like art—persists against all odds.

Within a week, the outside world was also looking up—one of her stories was accepted in *The London Magazine.* She wrote in her journal:

> *My optimism rises. No longer do I ask the*
> *impossible. I am happy with smaller things, and*
> *perhaps that is a sign, a clue . . . Every day*
> *is a renewed prayer that the god exists, that he*
> *will visit with increased force and clarity.*

It was a conflicted clarity. She had a series of restless nights full of tortured dreams about her mother, about "old shames and guilts." She and Ted were about to move to London—a prospect that had filled her with anxiety, like all major change, but now she began feeling an "odd elation" at the thought of turning a new leaf.

On a windy mid-November day of gray but balmy weather, she took a walk with Ted under the open sky and the bare trees, listening to the last leaves rustle in the wind, watching a scarecrow in a cornfield wave its hollow arms, noticing the blackbird on the branch, the fox prints and deer tracks in the sandy trail, the blue-purple hills and the green underbed of the lakes, the mole hills and tunnels webbing the grassland. Something began stirring—some restive creative vitality that needed an outlet. She recorded:

> *Wrote an exercise on mushrooms yesterday*
> *which Ted likes. And I do too. My absolute*
> *lack of judgment when I've written something:*
> *whether it's trash or genius.*

That exercise became her poem "Mushrooms"—a quietly mischievous work of genius, paying homage to the indomitable nature of the creative spirit. By the following summer, it was on the pages of *Harper's*, marking a bold departure from Plath's previous work.

It is both a hope and a heartache to consider that, today, mushroom species from the genus *Psilocybe* are being used in clinical trials to effectively allay treatment-resistant depression—a breakthrough she never lived to see that might have saved her life.

MUSHROOMS

SYLVIA PLATH

Overnight, very
Whitely, discreetly,
Very quietly

Our toes, our noses
Take hold on the loam,
Acquire the air.

Nobody sees us,
Stops us, betrays us;
The small grains make room.

Soft fists insist on
Heaving the needles,
The leafy bedding,

Even the paving.
Our hammers, our rams,
Earless and eyeless,

Perfectly voiceless,
Widen the crannies,
Shoulder through holes. We

Diet on water,
On crumbs of shadow,
Bland-mannered, asking

Little or nothing.
So many of us!
So many of us!

We are shelves, we are
Tables, we are meek,
We are edible,

Nudgers and shovers
In spite of ourselves.
Our kind multiplies:

We shall by morning
Inherit the earth.
Our foot's in the door.

Singularity Squared

TWO YEARS after Marie Howe wrote her splendid "Singularity" to premiere it at the second annual Universe in Verse, a virus humbled humanity into remembering our delicate interdependence, hurling the world into lockdown, the soul into sorrow and confusion. In those early days of terror and disorientation, preparing a virtual edition of the Universe in Verse as a lifeline of sanity extended to the world, I had the idea of turning Marie's benediction of a poem into an animated short film. Ninety thousand people around the world tuned in from their safe havens, together in isolation, together for consolation.

Over the weeks that followed, the film traveled far and wide on the wings of the poem's uncommon tenderness for life, eventually reaching a gifted young poet in New York City, inspiring her to compose a response to it—an exquisite ode to our primeval bond with one another and the rest of nature, embedded in which is the intergenerational chain-link of inspiration from which all art is born.

SINGULARITY

(after Marie Howe)

MARISSA DAVIS

in the wordless beginning
iguana & myrrh
magma & reef   ghost moth
& the cordyceps tickling its nerves
& cedar & archipelago & anemone
dodo bird & cardinal waiting for its red
ocean salt & crude oil   now black
muck now most naïve fumbling plankton
every egg clutched in the copycat soft
of me unwomaned unraced
unsexed   as the ecstatic prokaryote
that would rage my uncle's blood
or the bacterium that will widow
your eldest daughter's eldest son
my uncle, her son   our mammoth sun
& her uncountable siblings   & dust mite & peat
apatosaurus & nile river
& maple green & nude & chill-blushed &
yeasty keratined bug-gutted i & you
spleen & femur seven-year refreshed
seven-year shedding & taking & being this dust
& my children & your children
& their children & the children

of the black bears & gladiolus & pink florida grapefruit
here not allied but the same   perpetual breath
held fast to each other as each other's own skin
cold-dormant & rotting & birthing & being born
in the olympus   of the smallest
possible once before once

◆

◆

◆

ACKNOWLEDGMENTS

My gratitude to Emily Levine, who opened my world
to poetry; to Hannah Fries, for the symphonic vision
that made this book possible; to Ofra Amit, for creating
such breathtaking beauty amid a raging war; to Lili Taylor,
for lending the loveliest of voices to the universe; to the
generations of scientists and poets who devoted their lives
to reverencing reality, helping us see the exquisite cohesion
of the universe and the universe we are.

*The mission of Storey Publishing is to serve our customers by
publishing practical information that encourages
personal independence in harmony with the environment.*

◆　　◆　　◆

EDITED BY Hannah Fries
ART DIRECTION AND BOOK DESIGN BY Carolyn Eckert
ILLUSTRATIONS BY © Ofra Amit
   End sheets by Smith, Asa. *Smith's Illustrated astronomy, designed for the use of the
   public or common schools in the United States.* New York, D. Burgess & Co, 1855.
   www.loc.gov/item/43041124.

TEXT © 2024 by Maria Popova

◆   ◆   ◆

•   •   •

Storey books may be purchased in bulk for business, educational, or promotional use. Special editions or book excerpts can also be created to specification. For details, please contact your local bookseller or the Hachette Book Group Special Markets Department at special.markets@hbgusa.com.

•   •   •

Storey Publishing
210 MASS MoCA Way
North Adams, MA 01247
storey.com

Storey Publishing is an imprint of Workman Publishing, a division of Hachette Book Group, Inc., 1290 Avenue of the Americas, New York, NY 10104. The Storey Publishing name and logo are registered trademarks of Hachette Book Group, Inc.

ISBNs: 978-1-63586-883-8 (hardcover); 978-1-63586-884-5 (ebook); 978-1-66864-498-0 (audiobook)

Printed in China through Asia Pacific Offset on paper from responsible sources
10 9 8 7 6 5 4 3 2

Library of Congress Cataloging-in-Publication Data on file

MARIA POPOVA thinks and writes about our search for meaning—sometimes through science and philosophy, sometimes through poetry and children's books, always through the lens of wonder. She is the creator of *The Marginalian*, born in 2006 under the outgrown name *Brain Pickings* and included in the Library of Congress permanent digital archive of culturally valuable materials, author of *Figuring* and *The Snail with the Right Heart*, and maker of the Universe in Verse— a charitable celebration of the wonder of reality through stories of science winged with poetry.

◆　◆　◆

OFRA AMIT is an award-winning illustrator who lives and works in Tel Aviv. She is a graduate of Wizo Design Academy in Haifa, Israel, where she studied visual arts.